GĦAR DALAM
THE CAVE, THE MUSEUM,
AND THE GARDEN
Birżebbuġa

NADIA FABRI

PHOTOGRAPHY
DANIEL CILIA

HERITAGE BOOKS
IN ASSOCIATION WITH

H Heritage Malta
2007

HOW TO GET TO GĦAR DALAM CAVE AND MUSEUM

By *public transport*:
Bus nos. 11, 12, 13 from Valletta main bus terminus and stop near St George's Bay, Birżebbuġa.

By *car*:
Main roads leading to Marsaxlokk and Birżebbuġa. Follow road signs to Għar Dalam.

GĦAR DALAM ROAD

GĦAR DALAM ROAD

GĦAR DALAM

BORG IN-NADUR TEMPLES

BIRŻEBBUĠA

Għar Dalam Cave and Museum
Birżebbuġa Road
Birżebbuġa BBG9014

Tel: 2167 7419
www.heritagemalta.org

Insight Heritage Guides Series No: 14
General Editor: Louis J. Scerri

Published by Heritage Books, a subsidiary of
Midsea Books Ltd, Carmelites Street,
Sta Venera HMR 11, Malta
sales@midseabooks.com

*Insight Heritage Guides is a series of books intended
to give an insight into aspects and sites of Malta's
rich heritage, culture, and traditions.*

Produced by Mizzi Design & Graphic Services
Printed by Gutenberg Press

Copyright © Heritage Books
Photography © Daniel Cilia

First published 2007

ISBN: 978-99932-7-144-4

INTRODUCTION

Ghar Dalam is one of Malta's most important national monuments since it contributed a great deal to our knowledge on life on the Maltese islands in the remote and in the more recent past.

The main aim of this guide book is to explain Ghar Dalam's significance and importance. It is a naturally water-worn cave on the south-eastern part of Malta, about 500m away from St George's Bay, lying at the side of Wied Dalam. It is about 18.5m above sea level and is some 144m long. The first 50m are accessible to the general public. Access beyond this point is strictly forbidden for safety reasons. Beyond this point the cave starts narrowing; in some areas it becomes very low and narrow and the floor of the cave gets more and more uneven. In the lateral chambers beyond the first 85m there is perpetual darkness.

Ghar Dalam is the only cave in Malta where one can study the Pleistocene or the Ice Age fauna in an uninterrupted sequence dating back to 180,000 years ago having the latter all in one place.

THE CAVE

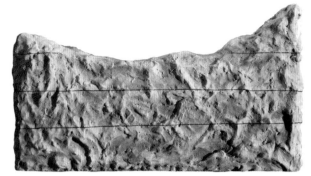

Geology, formation, and features

The Maltese islands consist of five geological layers lying horizontally one on top of the other in a 'cake-like' formation. The stratigraphy of the islands, starting from the uppermost layer, is the following: Upper Coralline Limestone, Greensand, Blue Clay, Globigerina Limestone, and Lower Coralline Limestone. All the Ghar Dalam deposits rest on the lowermost rock formation, the Lower Coralline Limestone formation.

The Maltese geological stratigraphy is nearly all made up of sedimentary

MALTA AND GHAR DALAM THROUGH GEOLOGICAL TIME
INTEGRATED INFORMATION

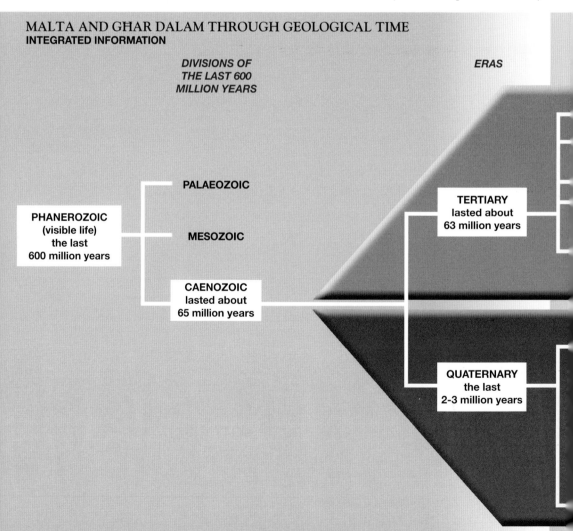

DIVISIONS OF THE LAST 600 MILLION YEARS

ERAS

PALAEOZOIC

PHANEROZOIC
(visible life)
the last
600 million years

MESOZOIC

CAENOZOIC
lasted about
65 million years

TERTIARY
lasted about
63 million years

QUATERNARY
the last
2-3 million years

rock. Most of the sedimentary rocks have at least 80 per cent calcium carbonate in them. When the calcium carbonate gets in contact with water, it starts dissolving. This reaction is called *chemical erosion*. Apart from this, mechanical erosion resulting from wind and pebbles in the area further erodes the limestone. This process eventually led to formation of the cave as we know it today.

Caves develop generally along lines of weaknesses within stratigraphic layers. Two types of caves exist: a vadose cave and a phreatic cave. A vadose cave can be found above the water table and is visible, while a phreatic cave is formed below the

Stage 2   Cavity formation due to perculating water through the rocks

EVENTS		RESULTS
Deposition of Marine Sediments	▶	With time and pressure, the marine sediments hardened into rocks
European and African Plates collide	▶	Upraising of sea bottom to form Alps, Siculo-Tunisian Ridge Birth of the Maltese islands exposing the local rock sequence
Abundant Rain = Floods + River	▶	• Rock surface erosion (Wied Dalam formation) • Waters carry away animals, soil, and pebbles
Forceful penetration of water through cracks and parallel bedding planes	▶	• Formation of underground Solution Tunnel at right angles to overlying river • Deposition of the insoluble residue (Bone-Free Clay Layer)
Further valley cutting with: Penetration of Subterranean Tunnel Splitting of Tunnel in two	▶	• Roof collapse • Discharge of River Load • Deposition of Hippopotamus Layer, Pebble Bed, Deer Layer • Formation of G'ar Dalam on one side and Second Cave on the other side of valley • River Bed lower than cave entrance • Cave Dry
G'ar Dalam used by man since c.5200 BC	▶	Deposition of Cultural Layers

Lowering of the valley bed

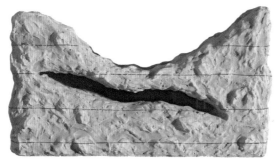

Ceiling of tunnel collapses giving rise to today's topography

Ghar Dalam

water table. Għar Dalam is a typical example of a phreatic cave. Phreatic is derived from the Greek word *phrear, -atos* literally meaning 'a well'.

It is important to note that during that period of time which coincides with the Ice Age, about 1.8 million years ago, the topography of the area was much different from today. The present-day valley was not yet formed and the area was basically flat. As a result of inclement weather during that time, rain water percolated through the soft rocks, dissolving them in the process and forming a cavity within the rocks at right angles underneath the flowing river. This cavity gradually became bigger and formed an underground tunnel. Continuous erosion from the river put a large pressure on the ceiling of the tunnel, causing it to collapse

Opposite page: Stalagmites and stalactites in Għar Dalam cave

Wall and ceiling fissure-orientated conical structures. Note ribbing formation

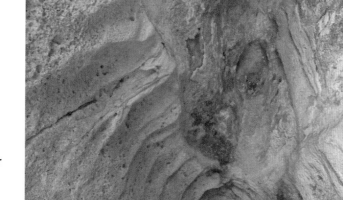

and depositing with it soil, animal bones, pebbles, and other material at the floor of the tunnel. This process gave rise to the new topography we see today: the Dalam valley and the two ends of the tunnel, Għar Dalam cave on one side and a much smaller shallower cave on the opposite side. Scientists have come up with the theory of the tunnel because of the fact that the same sediments were found within the two caves found on opposite sides.

Today Għar Dalam is considered to be one of the oldest caves in the Maltese islands. Evidence of its remote antiquity is its stalagmites and stalactites, which take an extremely long time to form. Stalactites form by the dripping of water from the ceiling of a cave. On contact with limestone, much of the calcium within the rock bubbles out in the air leaving only lime behind. Some of the lime from the ceiling may also drop on the ground beneath through the action of gravity. This leads to the formation of a stalagmite. The rate of growth of stalagmites and stalactites depends on the abundance of water-flow in the area. Eventually stalactites and stalagmites may join to form pillars.

Another interesting feature is the 'ribbing' effect that one can notice upon entering the cave. It is found on the left-hand side overhead the ceiling of the cave. This is the result of the effect of water working its way through the rock giving rise to a feature that this similar to a human being's rib cage.

On entering the cave one notices large dark patches on the walls. This dark mat-like formation is a living primitive floral species, algae. Inside Għar Dalam two different species of algae have been identified: *Anacystis montana* and *Schizothrix calcicola*. Artificial lights also encourage the growth of these algae in the middle parts of the cave. One can see them thrive on one of the large stalactites which is illuminated by artificial light.

Spanish sparrow nest

The present-day cave fauna

Caves and underground habitats provide a unique ecosystem for a variety of animal life forms. In most cases one finds a symbiosis between these different creatures, one depending on another. If one species is 'removed' from the cave, others will suffer the consequences. Għar Dalam is known to host over forty different species of life forms, from anthropods (include crustaceans, insects, spiders, and relatives) to birds to mammals. Different life forms require different habitats and the creatures present in Għar Dalam vary. They consist of those living in the lighted zone at the threshold of the cave such as birds and some insects, those living in semi-darkness half-way through the cave such as bats, and those living in complete or perpetual darkness such as woodlice. The most evident life forms are Spanish Sparrows *Passer hispaniolensis*. These common resident birds build their nests on ledges and in crevices in the outer part of the cave. Formerly the cave also used to host a large colony of the Mediterranean Mouse-Eared Bat *Myotis punicus*, but the only representative of these flying mammals are single individuals of Grey Long-Eared Bats *Plecotus austriacus*. Regular as a clock, February sees the emergence of hundreds of harmless burrowing bees *Anthophora acervorum*. These industrious insects lay a single egg in the loose soil. In turn, a parasitic wasp *Ronisia barbara* lays its egg inside the larva of the bee.

Ghar Dalam is also the type locality of a specialized endemic arthropod, known to the scientific world as *Armadillidium ghardalamensis*. This tiny woodlouse was discovered in 1982 by Prof. E. Caruso from the University of Catania and Carmel Hili, a Maltese student studying this group of animals. It was found that this particular woodlouse, which spends its entire life cycle in the darkness of the cave, evolved into a distinct and unique species. Its relatives are usually different shades of grey in colour, but *A. ghardalamensis* is white. It has also lost completely its sense of sight, and, as a defence mechanism, it adheres to the cave walls, a habit usually associated with limpets, whereas the other members of this group usually roll into a ball. It is highly sensitive to light and this is why the inner part of the cave is kept in darkness.

Because of this woodlouse, Ghar Dalam became a Special Area of Conservation (SAC) within the European Union Natura 2000 Framework.

CHRONOLOGY OF MALTESE PREHISTORY
BASED ON CALIBRATED CARBON 14 DATES*

Period	Phase	Date
Early NEOLITHIC		
Ghar Dalam	Ghar Dalam	5200 – 4500
Grey Skorba	Grey Skorba	4500 – 4400
Red Skorba	Red Skorba	4400 – 4100
LATER NEOLITHIC		
Żebbuġ	Żebbuġ	4100 – 3800
Mġarr	Mġarr	3800 – 3600
Ġgantija	Ġgantija	3600 – 3000
Saflieni	Saflieni	3300 – 3000
Tarxien	Tarxien	3000 – 2500
BRONZE AGE		
Tarxien Cemetery	Tarxien Cemetery	2500 – 1500
Borg in-Nadur	Borġ in-Nadur	1500 - ?
Bahrija	Baħrija	900 – 8th century BC

* Rounded to nearest calibrated Radiocarbon dates

First recorded human settlement

Maltese archaeology is divided into different phases and the earliest phase is known as the Għar Dalam phase. This is due to the fact that the oldest pottery was unearthed from this site, and to date these artefacts provide the only concrete evidence of early humans in Malta. This first farming-based society arrived on rafts from Sicily about 7,500 years ago. Their provenance is evidenced from the impressed type pottery they brought with them, which has a close affinity with the pottery found at Stentinello in south-eastern Sicily.

Old wall in 1912

Human presence and influence inside the cave

Apart from 'storing' the natural deposits of bones, the cave also served other purposes through the years, mainly concerning human activity.

Present-day entrance showing the triangular rock

Għar Dalam as a cattle pen

Caves have served as shelter to man and wild animals since prehistoric times. Man made use of them as his first abode and later on utilized them as cattle pens. Għar Dalam is one such cave. Apart from pottery, human

remains were also found inside Għar Dalam. Middens (ancient rubbish pits) have revealed animal bones and other remains in and around the cave. In fact, the cave served as a cattle pen right up till the earliest excavations in the second half of the nineteenth century. Farmers erected two rubble walls for the purpose. The first was erected at the entrance of the cave and the other some 80m inside to prevent animals from straying into its dark passages. Although the cavern has one very obvious entrance hole, it is not excluded that some narrow crevices may join up with another cave farther down the road (Għar il-Friefet). Recent studies, although not conclusive, have shown the presence of an endemic cavernicole (cave dweller) animal, previously known only from Għar Dalam, as present also inside Għar il-Friefet.

Evidence of this cattle pen can still be observed just outside the cave. A triangular rock with a circular perforation on one side, which is today surrounded by a chain, was used to tether animals, usually a male goat or a ram, as it browsed on the low grass or fodder. Inside, on the left wall by the path, there is another hole burrowed in the wall (belay) used for the same purpose. The rubble walls were pulled down in 1912 and the cave was not used for the rearing of animals anymore.

Shelter for Second World War refugees and an aviation fuel storage depot

A somewhat lesser-known aspect of the history of the cave is its use as a war-time shelter. Some 200 persons lived in the cave during August and September 1940. The refugees remained in the cave until October, when the

British military took over. They were compelled to evacuate the cave and find alternative lodgings, as the Royal Air Force wanted to use Għar Dalam as a storage depot for aviation fuel. They also filled the cave with sandbags and debris to level the cave floor. The cave and the museum remained closed for public viewing until 1947.

The graffiti

Along the cave walls, there are several scratched initials, dates, and other markings. Further inside the cave one meets with the initials GD, which stand for Giuseppe Despott. Archaeologists used to leave their mark in those sites where they had carried out excavations. This habit of writing one's initials was inherited from the Italian archaeologist Giovanni Belzoni who, while employed by the British Museum, ruined entire tomb walls in Egypt with his name and date of discovery.

Giuseppe Despott scratched marking

Napoleon Tagliaferro outside Burmegħeż cave

Arturo Issel

Further inside the cave on the left-hand side there is a slab of limestone with the date I.VI.XXII (1.6.1922) inscribed. This is the initial day when Napoleon Tagliaferro excavated in Għar Dalam. Another date, 01.01.1928 was made by G. Sinclair. There are also numerous illegible signs dating back to the time when the cave was used as a fuel depot.

**Right:
Shark's tooth**

Scientific recognition and excavations

The first recorded reference to Għar Dalam is that by Giovanni Francesco Abela in *Della Descrittione di Malta* (Malta, 1647), but its interesting remains were still unknown then. The scientific importance of the cave was not recognized until the latter half of the nineteenth century. Until then little, if anything, was known about its treasures. An indication of its importance came to light in 1865 when Arturo Issel (1842-1922), a Genoese geologist, came to Malta in search of remains of Palaeolithic Man. Issel started his excavations inside Għar Dalam by digging a shallow trench, some 60cm deep, to be able to investigate a specific area. Instead of finding remains pertaining to Palaeolithic Man, to his surprise Issel encountered hippopotamus bones and a number of potsherds. The latter were later identified as an 'impressed ware' type pottery, some of which

Gate installed at mouth of cave

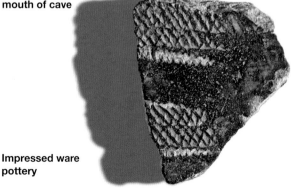

Impressed ware pottery

impressed with a shark's tooth, dating back to the earliest phase of Maltese prehistory, around 5000 BC, which later became known as 'Ghar Dalam Phase Pottery' after the place they were found in.

After Issel's discovery, the cave started to be visited by other scientists. However, not all of them had genuine intentions, and for years poachers raided the bone deposits. This depredation was only stopped when a gate was installed at the mouth of the cave.

John Henry Cooke, a teacher of English at the Valletta Lyceum, carried out the first serious work seventeen years after Issel. He excavated inside the cave from 1892 to 1893. Cooke dug out eight trenches at regular intervals of various sizes and depths, the first trench being at a distance of about 9.14m and the last being about 121.8m from the entrance.[1] This was one of the most important excavations since a large number of hippopotamus bones and remains of deer and elephants were discovered, as well as a complete jaw of a brown bear. This latter find represented a concrete evidence of carnivorous animals during the Quaternary Period about 1.8 million years ago on the Maltese islands. Human implements were also discovered for the first time during this excavation.

Following Cooke, there was a pause in excavations for about two decades. In the meantime, however, Cooke continued to investigate Pleistocene deposits in Dwejra and Wied il-Ghasri, Gozo, from where similar Pleistocene fauna was retrieved. His findings stimulated various other scientists who carried out other important excavations.

In 1912-13, Napoleon Tagliaferro, rector of the University and

Flint tools and obsidian

Left:
Needles made out of animal bones

Overleaf:
General view of G'ar Dalam

John Henry Cooke

archaeologist, together with Giuseppe Despott, who later became the first curator of the Natural History Section of the Museums Department, dug a trench about 121.8m from the entrance. This excavation was taken over by the British Association a year later. Despott, who now coordinated the works with Themistocles Zammit, the curator of the Valletta Museum, and Dr Thomas Ashby, an archaeologist, cut another trench in 1914 at about 61.7m from the entrance. Much of the material found in this excavation was eventually transferred to the small museum which was opened on site in 1932.[2]

In 1917 Despott received a grant of £10 from the British Association to resume his work. He conducted excavations up to 1920, mainly on the left-hand side of the cavern, some 24.4m from the entrance. This excavation had another significant find. Apart from other animal remains, a skull of an elephant was found, possibly belonging to *Elephas*

melitensis, (*Elephas melitensis* is now considered as a synonym of *Elephas mnaidriensis*), with its cervical vertebrae in situ. These deposits were found at a much lower level from the 'deer layer' and were left *in situ* in order to show their exact position.[3] Several tools and weapons were also unearthed, together with pottery with impressed or incised decorations before being baked.

In 1922 Mr G.G. Sinclair, a civil engineer with the British Admiralty, and Mr E. Flamingo, who investigated the physical aspect of the cave, excavated this trench once again.[4] Ever since this excavation, the sides of this trench started to deteriorate. In order to retard or even possibly avoid the process of deterioration, two stone pillars were constructed to support the portions of the trench still in place. However, the trench was not properly covered and it collapsed owing to large quantities of rain in 1926, burying all the material excavated and purposely left *in situ* by Despott in 1917.[5]

Għar Dalam right-hand wall

This trench revealed that the floor was actually made up of great side wall slabs that had collapsed inwards on an ancient clay deposit during the early stages of the formation of the cave. This left a central trench into which most of the remains had been washed. Another 4.2m deep trench was excavated with the hope of finding the original rock, however *torba* (pressed earth material) was found and this excavation was abandoned. A further trench of about 25.9m was dug outside the entrance at a lower level. Here the rock bottom proved to be the same and *torba* was found again.

Ms Gertrude Caton Thomson, a British palaeontologist, initiated the 1922 excavations, which were later continued by Giuseppe Despott, George Sinclair, Carmelo Rizzo, and J.G. Baldacchino, among others. Most of the material found during these excavations represented a storage problem for the Museums Department, even if the preservation of retrieved remains was being taken care as much as possible on site.[6] Despott, in fact, also constructed a small room a few metres away from the cave in which to store and clean the retrieved material. However, the storage problem was so acute that for some time excavations had to stop. The most important examples were preserved in Birzebbuga since 1914, while a good number of semi-fossilized bones have been lying at the entrance of the cave since 1931.[7] However, between 1931 and 1932, a large quantity of the specimen material was transferred to the Museum at the Auberge d'Italie in Valletta. Most of this material had been obtained from the excavations conducted by Ashby and Despott in 1914.

Following the untimely death of Despott in 1933, his post as

Joseph G. Baldacchino upon the discovery of an elephant tusk

curator was taken by Dr Joseph G. Baldacchino, who dedicated much of his time to unearth bones from Għar Dalam and to try to solve some of its riddles. He also set up and organized the 'old museum', which is still in its original state. Between 1933 and 1938, he unearthed numerous organic remains, including those of bears, wolves, foxes, deer, shrews, and dormice, as well as two large elephant tusks. Ms Dorothea M.A. Bate of the Natural History Museum of London was of instrumental help to Baldacchino since she examined the vertebrate remains which he unearthed from the cave. In 1937 Baldacchino published a pamphlet in which the theory was expressed for the first time that '*the two caves*

Dorothea M.A. Bate

on opposite sides of Wied Dalam once formed one continuous tunnel which was gradually cut through forming today's valley with the caves on either side'.

Excavations came to a standstill at the outbreak of the Second World War when both cave and museum were closed to the public and were only re-opened in 1947. Twenty-five years later, the cave received further attention when, in 1974, a team led by Dr Gerhard Storch of the Senckenberg Musem of Frankfurt, Germany carried out a series of studies on the micro-mammals and avian fauna in the Għar Dalam deposits. Remains of bats, shrews, rodents, and birds were discovered and identified.

Neanderthal Man in Malta?
The Taurodont teeth and their controversy

A breakthrough in the excavations was announced in the summer of

1917, when two taurodont molars, possibly from two different humans, were discovered among several deer teeth in the Deer Layer in the second trench. Despott continued to investigate and he encountered more human teeth. Taurodontism in humanoids had been previously described by the renowned anatomist Sir Arthur Keith a few years earlier, when he attributed this dental feature exclusively to Neanderthal Man. Incidentally, it was Keith who had coined the term 'Taurodont', meaning 'bull tooth'. The discovery led Despott to speculate that this species of humanoid might have existed in Malta during the mid-palaeolithic (200,000-100,000 years ago). Since Neanderthal man had a strong affinity with caves, it was considered worthwhile to investigate the possibility of further evidence inside Għar Dalam, and a possible third molar was unearthed during the 1936 excavations.[8] Debates on these

Għar Dalam cave in 1921

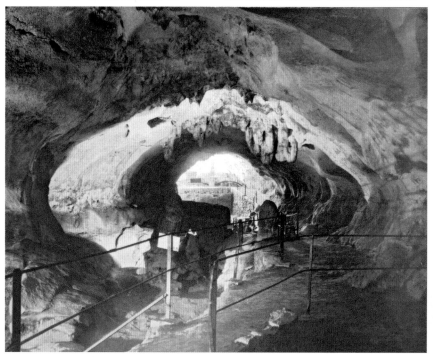

Sir Arthur Keith

findings ensued and championed by none other than Sir Arthur Keith himself. On inspecting the molars, Keith commented that: 'They are hard, heavy and mineralized; the enamel is of bluish dark opalescence; the neck and root are of a dull chalky grey … The enamel of the cusps is sharp and crystalline … the details of cusp formation differs from those seen in the cusps of modern man, particularly as regards the size and length of the postero-internal cusps. In size and form such teeth have been seen in no race of mankind *except H. Neanderthalis*; in condition of fossilization and in the fauna which keep them company; in the red cave earth in Għar Dalam, they are in their proper Pleistocene setting.'

Keith therefore prematurely concluded that these molars belonged to Neanderthal Man and went on to give his explanation on how the animal bones arrived in Għar Dalam, stating that Neanderthal Man rounded up the animals inside the cave to slaughter them.

The discovery of a human molar in 1917 by labourers excavating inside Għar Dalam on behalf of the curator sparked a series of controversies which still linger on today. Excavating bones and fossils in the early twentieth century was far from a meticulous

**Right:
A 1923 photograph**

Human taurodomt teeth

process. Dental tools, trowels, toothbrushes, and hairbrushes are the standard field tools of today, whereas in those early days pickaxes and shovels were used. Moreover, illumination inside the cave was provided with the naked flame, and untrained personnel carried out the actual digging. In fact Despott had a number of quarry workers digging for him inside Għar Dalam. Therefore it is possible that these teeth fell down from upper levels and settled on the exposed Deer Layer where they were found. To add weight to the fact that taurodontism is not exclusive to Neanderthal Man, in 1962 Professor J.J. Mangion, a Maltese dental surgeon, extracted a taurodont molar from a Maltese man from Zebbug. In 1963 Dr Kenneth Oakley, the same person who some

J.J. Mangion

Tools used to sharpen arrow heads

K.P. Oakley

years earlier had exposed the Piltdown Man forgery,[9] carried out a series of tests on the Għar Dalam molars and concluded that they are not older than the Neolithic, around 5000 BC. Two Maltese medical doctors recently stated that the latter results were tampered with and have contested their validity. However, the 'facts' they presented lack concrete scientific basis. One has to be very careful when trying to interpret data from such scanty and dubious contexts. It is highly probable that a final and definite answer will never be found.

The robbery of 1980 and the last excavation

The night of 7 April 1980 is considered as a black day in Maltese palaeontology. Unknown persons broke into the Għar Dalam museum and carried away several specimens, including the two tusks of the Maltese

dwarf elephant discovered by Dr Baldacchino in the 1930s. Despite many investigations at the time of the robbery, the thieves were never caught. Scientists and locals are only left with a photographic record of this elephant.

Dr George Zammit Maempel, then the curator of Għar Dalam, assisted by J.J. Borg, carried out the last excavation in the cave in November 1995. Remains of a large-sized hippopotamus were found. However, because of the friability of the bones, only the lower jaw, with part of the dentition still in place, and a pelvis, possibly belonging to the same specimen, were extracted. They are today displayed

Pelvis and lower jaw of a hippopotamous

**Opposite page:
The hippo layer in
the middle trench**

in the new museum wing in showcase Number 16.

The Ice Age

In order to understand the faunal aspect of Għar Dalam, one has to first understand what was going on globally, especially in Europe, in the Pleistocene Epoch. In this era, around two million years ago, and which is also the earliest division of the Quaternary Period, the climate changed constantly and rapidly, giving rise to glacial cold stadials and temperate periods. This means that climate went through periods of thawing or 'interglacials' and glacial periods (or cold and warm climate) for numerous times. The most feasible theory to date about these climatic fluctuations is based on the idea of the Yugoslav mathematician and astronomer Milutin Milanchovitch, and is commonly known as the Milanchovitch factors. He observed the following three variations: (a) the shape of the earth's orbit around the sun, (b) the tilt of the earth's axis, and (c) the time of year that the earth is closest to the sun. All of these three factors varied during the Pleistocene Period and contributed to this fluctuation in climate. The Holocene, the most recent epoch of the world's history, comprises the last 10,000 years of warm climate.

This period of time, the Pleistocene, is commonly known as the 'Ice Age' since during that time, rivers and most of the sea turned into glaciers. These ice sheets covered most of central Europe and the northern hemisphere. The sea-bed between Malta and Africa before and after the Ice Age was about 1,100m, while that between Malta and Sicily was about 90m. Owing to ice retreat during the ice age, the sea level in the Mediterranean dropped by 250m, giving rise to new land-bridges connecting countries together. This was the case between Malta and Sicily. Since land was uncovered, animals could roam freely from Sicily to Malta. However the sea level between Malta and Africa remained too deep for animals to swim all the way from Africa to Malta or vice versa. This should help clear the misconception which is still widespread today that during the Ice Age animals travelled from Africa to Malta. Although it is true

**Life on earth in
the last 600 million
years**

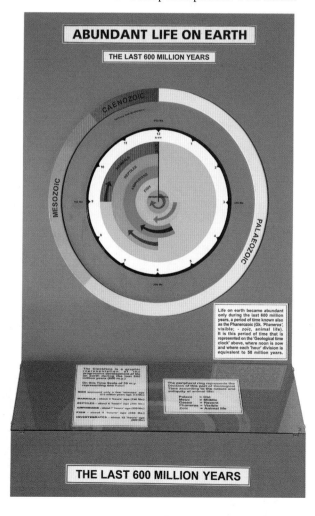

THE LAST 600 MILLION YEARS

Hippo bones in situ

that geologically Malta is connected to Africa, the sea barrier renders us completely different from each other.

Effects of the Ice Age on the Maltese islands

Unlike Europe, Malta did not experience glaciation. Therefore, instead of an Ice Age, Malta experienced a Rain Age or Pluvial Age. However, Malta still went through a number of changes concerning:

1. **Climate** – torrential rains and floods. This resulted in sweeping away living animals in Malta with the action of currents, and the carving of valleys. Most of the valleys still visible in the Maltese landscape today, including Wied Dalam, are the result of this period of time.
2. **Sea level** – fall in sea level which joined Malta to Sicily.
3. **Fauna** – arrival of exotic fauna, such as elephants, hippopotamus, bear, wolf, etc… that travelled to Malta due to an 'extended' landmass.
4. **Animals** experienced an evolutionary change to ensure their survival.

The Pleistocene fauna: Adapting to a changing environment

Adaptation is the way a living organisim copes with environmental stresses and pressures that are subject to it. Some organisims are able to cope and therefore they adapt and evolve, while others would not cope and, as a result, they become extinct. The animals that had travelled south by means of the land bridge between Malta and Sicily and were trapped in the Maltese islands when the sea level rose again were faced with this problem. This process was not sudden but took thousands of years when the ice sheets retreated again. The small size of the island could not possibly provide enough food to support herds of large-sized animals. This is one of the reasons why these animals had to undergo evolutionary changes in order to adapt themselves to the island's conditions.

Animal populations living in isolation frequently experience changes in stature which generally results in stunting (dwarfing) or gigantism to ensure their survival. These physiological changes take place owing to an accumulation of events over time and are considered to be a slow and progressive evolutionary process. The animal has two options: either to die or else to adapt itself to a new environment. By nature, animals 'choose' to adapt themselves and evolve into either smaller or bigger versions, according to their needs. Such an evolutionary development was experienced not only in Malta but was also a characteristic of other Mediterranean islands, such as the Balearics, Sicily, Cyprus, and Crete. The degree of dwarfing or gigantism varies from island to island, but the Maltese Pleistocene fauna is very similar to that of nearby Sicily.

Animals require specific habitats to live comfortably, with their basic requirements being territory and mating. If many animals of a large size inhabit one particular area, space is going to be much reduced. Another reason for reduction in body size is the possible absence of predators. Thus, animals could afford to become become smaller since there is no need for a huge size to deter predators. A reduction in size also automatically means less food requirements, therefore giving the animal a much better chance of survival.

Together with Sicily, Malta also shares the phenomena of gigantism of its smaller Pleistocene mammals. Gigantism occurs mainly when the animal is prolific, meaning that it breeds continuously. The giant dormouse *Leithia melitensis* is an example of gigantism of the Pleistocene fauna. It was basically the size of today's guinea pig. In cases where animals breed a lot, there is a high competition for food as well as for partners. Thus they have to become bigger to 'win' both food and territory. The fact that animals like the giant dormouse grew in size does not reflect the absence of predators, since in some Quaternary sites in Malta remains of wolves, foxes, and raptors, including owls, were found alongside bones of dormice and other rodents. Another example of gigantism is that of the Maltese Pleistocene tortoise *Testudo robusta*, which reached a size nearly as big as today's Galapagos islands tortoises. This was also the case of the giant lizard *Lacerta wiedincitensis*, which is estimated to have been some 70cm long during the Pleistocene. A jaw of this lizard was unearthed by Dr George Zammit Maempel from Wied Incita, Attard in 1973.

It is possible that the size of the fauna present in the Maltese islands during the Pleistocene period had already undergone body changes before arriving to Malta. Recent studies indicate that the majority of dwarfing and gigantism occurred in Sicily or southern Italy. However, this issue of dwarfing and gigantism in Sicily and Southern Italy is still in debate by palaeontologists. Various studies on animal evolution have revealed that animal size does not depend on the size of the islands, but mostly on the duration of the isolation.

Organic deposits and layers

At a certain point in time during the late Pleistocene, the river carved its way deep into the limestone until it reached the underground solution cavity. Continuous pressure broke the ceiling of the tunnel, depositing soil, pebbles, stones, bone remains, and other material. This pile of material gradually spread laterally but it never reached the end of the tunnel. This is the reason why remains in Għar Dalam are limited to the outermost 80m.

The different layers inside Għar Dalam were deposited in different periods in time and have formed six main layers of deposits, generally labelled according to the main animal species or characteristic material which were encountered within that specific layer.

The six layers present in Għar Dalam are:

1. The **Bone-free Clay Layer** (c. 1.75 m) is the lowermost layer of the cave's stratigraphy. Before the ceiling of the tunnel collapsed, a thick deposit of clay had already been formed, which consisted mainly of insoluble limestone residue. As the second cave (the one opposite Għar Dalam) is at a lower level, it was completely filled with this clay deposit, leaving practically no place for the deposition of other layers that are found in Għar Dalam.

2. The **Hippopotamus Layer** (c. 1.7 m), dating to about 180,000 to 130,000 years ago, is the second layer and is found on top of the Bone-free Clay Layer. This layer is so called because the predominant animal remains are those of hippopotamus, even though elephant remains were also present. Since these remains had been rolled along the river bed, the edges and fine features of the bones have been eroded. With time, and the dripping of lime-rich water, the elephant and hippopotamus bones dried up and hardened into a solid mass known as bone breccia. These bones make up to about 75 per cent of this layer.

3. The **Pebble Layer** (c. 0.35 m) is the next layer. This is the stratum that separates the Hippopotamus Layer from the Deer Layer. The absence of animal bones in this layer is indicative of the scarcity of animal life in Malta.

4. The **Deer Layer** (c. 1.75m) contains animal remains belonging to the latter part of the Pleistocene (18,000-10,000 years ago). These remains are predominantly composed of antlers and of long bones of two species of deer. However, other remains pertaining to wolf, brown bear, and red fox were also present within this layer.

5. The **Calcareous Sheet** is a thin layer about 6cm thick that separates the Deer Layer from the Cultural Layer. This is a sterile layer corresponding to a volcanic ash layer discovered in open sites such as at Mrieħel.

6. The uppermost cave deposit is the **Cultural** or **Domestic Layer** (c. 0.75 m). This stratum is the product of the habitation of the cave by humans and domestic animals throughout the last 7,000 years, since the arrival of first man on Malta around 5,200 BC. Human and animal remains, as well as pottery, were also

6 DOMESTIC ANIMALS LAYER
(CULTURAL LAYER, POTTERY LAYER)
c. 7th c.

5 CALCAREOUS SHEET
c. 8 c.

4 DEER LAYER
c. 9 c.

Upper neck bone

Dear toe nail

present in this layer.

Having just seen the different layers in Għar Dalam, a look at the organic remains found within them is appropriate.

The Hippopotamus Layer includes various remains of large- and small-sized animals. Three species of hippopotamus (*hippo* = horse and *potamos* = river therefore river-horse) were identified, the large-sized *Hippopotamus amphibious*, and the two smaller *Hippopotamus pentlandi* and *Hippopotamus melitensis*. Elephants were represented by the following species: *Paleoloxodon mnaidriensis*, which was about 1.5m high at the shoulder, and *Paleoloxodon falconeri*, which was about 90cm at the shoulder. On the other end of the scale, toads, bats, shrews, rodents, and bird remains were also found.

The Deer Layer is characterized by the large quantity of deer remains belonging to an undersized form of the red deer *Cervus elaphus* and the fallow deer *Dama dama*. This layer is particular for yielding remains of carnivorous animals, namely a small-sized form of the brown bear *Ursus arctos*, a small-sized wolf *Canis lupus*, fox *Vulpes* sp., otter *Lutra euxena*, wild ass *Equus* sp., and bats, shrews, rodents, and birds.

The Cultural Layer is characterized by the abundance of domestic animal remains as well as human organic and inorganic remains. The remains of cows, sheep, pigS, cats, and other introduced animals were unearthed from this layer. Remains of bats, which probably were the only natural occurring species in this layer, were also found. This uppermost layer yielded human remains as well as stone implements and pottery from the earliest human communities in Malta dating back some 7,200 years ago.

Much of the bone deposits of the cave have now practically been excavated. The entire sequence is preserved in a sample stratigraphic column and wall found about 18m from the gate.

Fragment of bone breccia

THE MUSEUM

The old Victorian-style museum

As a result of a lack of storage for the bone remains that were unearthed in Għar Dalam, in 1929 Giuseppe Despott requested the government to erect a building on site to house all the findings discovered by himself and his predecessors. However, owing to ill health, Despott was compelled to retire in 1933. Therefore, he did not start work on the display and never had the opportunity to see this museum installed and set up. This work was actually undertaken by his successor, Joseph Baldacchino.

Baldacchino presented to the public a series of repetitive exhibits mounted on wooden boards displayed in Victorian style. The museum was inaugurated in two stages: the display in the corridor was inaugurated during the financial year 1934-35, while the main hall was unveiled during the financial year 1936-37. Typical of the age, rather than on quality,

Għar Dalam entrance

**Left:
Fragment skull of deer**

Baldacchino's display

SWABA' TAS-SAQAJN
2ⁿᵈ PHALANX (TOE BONES)

GHADAM TAS-SWABA'
DIGIT BONES

Toe bones

with genuine mounted skeletons of a brown bear *Ursus arctos*, a young elephant *Loxodonta africana*, a young hippopotamus *Hippopotamus amphibious*, the skull of an adult hippopotamus, a red fox *Vulpes vulpes*, a red deer *Cervus elaphus*, and a wolf *Canis lupus*. All these skeletons belong to modern-day animals and are exhibited for comparative and educational purposes only. For historical purposes, the 'Old Museum' has remained untouched ever since, except for some minor changes and the introduction of more informative labels and electric lighting.

The new didactic museum

A dream lasting over twenty years finally took shape in the 1990s. Dr Zammit Maempel's idea was to present a didactic permanent display for visitors to enjoy a better experience of the history of Għar Dalam and its overall significance. The new museum wing was inaugurated in 2002.

The new museum bears information on life on earth, the effects of the Ice Age on Malta,

the emphasis was on sheer quantity. A number of upright, tall showcases were arranged along the four walls of the museum and the corridor, each unit being fitted with several wooden boards to which were wired a number of bone remains. However, these display cases bore no information whatsoever. Informative captions were subsequently installed in each display case to help the visitor understand better. Today this wing is referred to as the 'Old Museum'.

Baldacchino decorated the centre of the room

Adult hippo skull (recent animal)

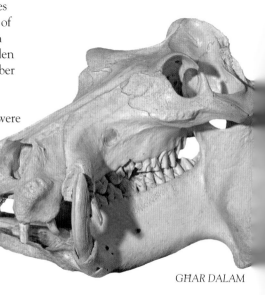

the formation of the cave, and dwarfing and gigantism in animals. It also presents information about hippopotamus, elephant, deer, wolf, fox, bear, and otter remains, and deals with the historical aspects of Għar Dalam. These include the first mention of the cave in 1647; early photos of the cave; reference to excavations and studies; and original publications by Thomas Spratt, Andrew Leith Adams, Hugh Falconer, Napoleon Tagliaferro, Dorothea Bate as well as by the three curators, Giuseppe Despott, Joseph Baldacchino, and George Zammit Maempel. One can also find information about the archaeological aspect of Għar Dalam. These are exhibited in the 'Għar Dalam Phase' display case where tools, amulets, and sling-stones can be seen. One can also find information about the controversy of the taurodont teeth.

Two large oil paintings by the Maltese artist Robert Caruana Dingli

Fragment of lower jaw of a deer

Fragment of a hippopotamus jaw

(1883-1940) decorate this room. The first represents a modern view of the countryside in the Għar Dalam area, while the other is an artist's impression of the all fauna recorded in Malta during the Pleistocene Period. These two paintings were commissioned to accompany the Malta Pavilion during the British Empire Exhibition held at Wembley in 1924.

The Didactic Museum

THE CURATORS

Giuseppe Despott, MBE (1878-1936) Born in Cospicua, Giuseppe Despott was educated at St Ignatius College, Flores College, the University of Malta, and the English Academies of England and Rome. He was the first curator of the Natural History Section of the Museums Department, and therefore the first curator of Għar Dalam, appointed in 1922. His fields of specialization included zoology and palaeontology. He was a member of various ornithological and zoological societies such as the Association of Marine Biology and the Conchological Society of Great Britain and Ireland.

His main contributions to Għar Dalam were that he unearthed numerous organic remains from the cave and was responsible for the construction of the building that was intended to house a museum, now referred to as the old museum, to display the remains extracted by him and his predecessors. However, because of his failing health, Despott was not able to set up the display. This task was later taken up by his successor, Dr Baldacchino. Despott was also far-sighted enough to preserve a sample of the cave deposits – the stratigraphic wall and column – for future generations. He achieved fame when he unearthed two taurodont molars from

the cave which sparked a controversy about the possible presence of Neanderthal man in Malta.[10] The letters GD in the cave stand for Giuseppe Despott; he etched his initials to mark the fact that from there he had unearthed various findings. A small shed was constructed outside the cave's mouth for the storage of tools and some retrieved material. Just before his early retirement in 1933, Għar Dalam cave was officially opened to the public.

Despott was a prolific writer. His publications include *Catalogue of the Collection of Birds of Malta* (1907), *The Ichtyology of Malta* (1919), *Excavations conducted at Gʻar Dalam, Malta* (1916, 1917, and 1923), *The Reptiles of the Maltese Islands* (1915), and *The Breeding Birds of Malta* (1916).

Dr Joseph G. Baldacchino, Ph. C., MD (1894-1973) Born at Siggiewi, Baldacchino studied at the University of Malta, from where he graduated as a pharmacist in 1915 and as a doctor in 1919. He served as a general practitioner for fourteen years but in 1933 he succeeded Giuseppe Despott as curator of Għar Dalam, originally as a part-timer, from 1933 to 1947, when he was appointed director of the Museums Department, a post he held until his retirement in 1955.

Baldacchino was a

dedicated person and had many achievements while curator. One of his earliest tasks as curator of Għar Dalam was the setting up of the museum's display, which today is still in its original state. Between 1933 and 1938 he unearthed numerous organic remains from Għar Dalam, including those of bears, wolves, foxes, deer, shrews, and dormice among other fauna, as well as two large elephant tusks. He also carried out several excavations in other Maltese Pleistocene deposits. In 1937 Dr Baldacchino proposed the idea that the two caves opposite each other of Wied Dalam once formed one single tunnel which was eroded centrally, thus forming today's valley. He passed away on 6 July 1973.

Dr George Zammit Maempel, MQR, D.Sc. (Hon. Causa), Ph, C., MD (1925-) George Zammit Maempel was born in Naxxar on 2 May 1925. He acquired a diploma in pharmacy in 1948 and graduated as a doctor in 1952. He also attended a course of geological studies and was awarded a UNESCO fellowship in museography in 1969.

Originally enrolled as a part-time assistant curator in 1967, Zammit Maempel has been the longest serving curator. In 1988 he abandoned his medical practice and devoted himself to palaeontological studies and museum curatorship and was appointed part-time curator.

He was responsible for the building of an additional museum wing which was inaugurated in 2002. Zammit Maempel also catalogued all the Gʻar Dalam material and discovered new pathological features in Maltese Pleistocene hippopotami and investigated the biology and ecology of the cave. He discovered a new genus and new species of a fossil dormouse, *Maltamys wiedincitensis*, during excavations at Wied Incita, and a new species of giant lizard about 70cm long, *Lacerta siculimelitensis*.

In 1992 Dr Zammit Maempel was honoured with the medal for distinguished service to the Republic, and in 1993 the University of Malta conferred on him the degree of Doctor of Science (Hon. Causa) as a reward for his studies on Maltese geology and palaeontology.

Zammit Maempel is author of *An Outline of Maltese Geology* (1977), *The Folklore of Maltese Fossils* (1989), *Għar Dalam Cave and Deposits* (1989), as well as a number of other scientific publications. He occupied the post of curator until 2003.

The current curator of Għar Dalam is **John J. Borg**, a field biologist by training who also studied the quaternary avian remains of Għar Dalam and various other sites.

THE GARDEN

The national plant of Malta: Maltese Rock-Centaury – *Widnet il-Baħar, Cheirolophus crassifolius*

This perennial wild plant is endemic to the Maltese islands and grows on cliffs in the southern part of Malta and Gozo, but it does not grow on Comino. It flowers mainly between May and July and, with favourable conditions, it may continue to blossom until September. The shape of its fleshy leaves is spoon-like, and the flower-head is violet in colour. It does not have any spines or bristles.

The Maltese botanist Stefano Zerafa first described it in 1827 when he also noted that it is endemic to the Maltese islands and named it *Centaurea spathulata*. Unfortunately this name had already been assigned to another plant, and the Italian botanist Bertoloni renamed it *Centaurea crassifolia* in 1829. When the Czech Botanist Josef Dostal studied it again in 1975, he found that the plant was a primitive plant and he therefore created a new genus – *Palaeocyanus*. In 1999, the genus was changed again to *cheirolophus* by Alfonso Susanna.

This plant is considered one of the rarest species in the world and is listed in the World Conservation Union (*Red Data List of Globally-Threatened Plants* – 1997 IUCN). Legally protected by legal notice 49/1993, it was declared the national plant of Malta in 1971.

Sandarac Gum tree

The national tree of Malta: Sandarac Gum Tree – *Siġra ta' l-Għargħar, Tetraclinis articulata*

The Sandarac Gum Tree was declared the national tree of Malta on 16 January 1992. This tree is of medium height and rarely exceeds 12m but usually grows up to 6m in the wild. It is pyramidal-shaped and has delicate branches covered with small scale-like leaves arranged in four rows. It grows male and female cones. The male cones are about 3mm wide while female cones are up to 15mm wide. The cones have five-pointed sides, which explains its Latin name since in Latin *tetra* means 'five' and *clinis* means 'points', therefore 'the five points of the cone'.

The tree grows in hot, dry regions, especially in maquis areas on slopes of Coralline Limestone. In Europe it occurs only in Malta and in Southern Spain but it can also be found in Morocco, Algeria, and Tunisia. In Malta it occurs in few localities including Il-Maqluba (limits of Qrendi) and Mellieha.

The Sandarac Gum Tree is very rare in Malta. According to the IUCN (International Union for Conservation of Nature), the Maltese populations of Sandarac Gum tree are listed as critically endangered.

The national bird of Malta: Blue Rock Thrush – Merill, *Monticola solitarius*

About 20.5cm in length, the Blue Rock Thrush is a resident bird of the Maltese islands. Its favoured habitat is the sea-cliffs and rubble screes, but it also inhabits rocky hillsides. Some can also be found in quarries and derelict buildings. Its distribution ranges from Gibraltar throughout the Mediterranean and eastwards as far as Japan. In March it builds its nest in a narrow crevice where two or four chicks are raised. A second brood is usually raised in June.

The melodious song of the Blue Rock Thrush echoes along the cliffs from February to May. In autumn, young birds can be seen perching on the roof of the watch tower and the gun post on the other side of Wied Dalam. The brown-coloured female appears drab when compared to the iridescence blue of the male. *Il-Merill* is depicted on the backside of the Lm1 coin and, like all the other breeding birds, it is legally protected.

Trees, bushes, and flowers in the garden

The Maltese islands, although small in size, boast over 1,000 different species of flowering plants, a wonderland for amateur and professional botanists. During the last few years a variety of

Aleppo pines and other flora

indigenous plants have been planted in the open spaces in front of the museum and along the footpath separating the museum from the cave.

Along the footpath one encounters a variety of indigenous trees and shrubs like the African Tamarisk (*Bruka*) *Tamarix africana*, the Pomegranate (*Rummiena*) *Punica granatum*, Oleander (*Difla*) *Nerium oleander*, Evergreen Oak (*Sigra tal-Ballut*) *Quercus ilex*, Olives (*Żebbuġa*) *Olea europea*, Myrtle (*Riħan*) *Myrtus communis*, Aleppo Pine (*Sigra taż-Żnuber*) *Pinus halepensis*, Judas Tree (*Sigra ta' Guda*) *Cercis siliquastrum*, Lentisk (*Deru*) *Pistacia lentiscus*, Rosemary (*Klin*) *Rosmarinus officinalis*, Common Sage (*Salvja*) *Salvia officinalis*, and the Bay Laurel (*Randa*) *Laurus nobilis*.

One of the most interesting is the Carob Tree (*ħarruba*) *Ceratonia siliqua*, an evergreen tree which can grow up to 10m in height. The male produces the flower, while the female tree produces the pod (the seed). Not all the trees have different sexes but the carob tree is found in the male and the female form. The tree was introduced in Malta in ancient times, probably from the Eastern Mediterranean, and has since become a common tree in the islands.

The carob seed was used as a weight measure (1 seed = 1 carat) in Phoenician and Roman times and is still a unit for measuring gold today; its pods were used as fodder for animals and also as food for humans in times of emergency such as during the Second World War. Rosary beads used to be made from its seeds, while a type of syrup or caramel is still produced from its flesh to be eaten during Lent (since it is derived from natural and not from artificial sugars) or else as popular medicine against coughs known as *ġulep*.

Added to these one can also find a few exotic plants that were planted earlier such as the Mock Orange (*Pittisporum*) *Pittosporum tobira*, Murray Red Gum (*Sigra tal-Gamiem*) *Eucalyptus camaldulensis*, Italian Cypress (*Cipressa*) *Cupressus sempervirens*, She-Oak (*Kaswarina*) *Casuarina strictam*, Hibiscus (*Hibiskus*) *Hibiscus rosa-sinensis*, Geraniums (*Sardinell*), Lavander (*Lavanda*) *Lavandula spica*, Jamaica Mountain Sage – (*Lantana*) *Lantana camara*, Agave *Agave americana*,

Olive tree

Wigandia

and Hottentot Fig *Carpobrotus acinaciformes*.

Għar Dalam is one of the few localities in Malta where the introduced Wigandia (*Wigandja*) *Wigandia caracasana* grows. Its large thick fleshy leaves and the violet flowers add a touch of colour to the area. One of these trees is found on the right-hand side on the front of the museum and another one grows on the side of the new museum.

The rocky grounds provide a sample of our wild plants such as the White Mustard *Diplotaxis erucoides*(*Gargir Abjad*), Fennel *Foeniculum vulgare* (*Bużbież*), Caper bush (*Kappara*) *Capparis orientalis*, Prickly Pear (*Bajtar tax-Xewk*) *Opuntia ficus-indica*, and Olive-leaved Buckthorn (*Żiju*) *Rhamnus oleoides*.

Walking down the footpath from the museum towards the cave, there are two green iron gates separated by a narrow path. In the early years of the museum, this was the old road leading down to St George's Bay.

Fauna present in the area

The garden and the nearby valley host a variety of animal life forms.

Pine cones

Insects are certainly the most common. Butterflies are represented among others by the Large White *Pieris brassicae*, the Small White *Pieris rapae*, and the Swallowtail *Papilio machaon melitensis*. Numerous species of beetles, bees, and wasps are attracted to the abundant flowers. The chameleon *Chamaeleo chamaeleon* is an introduced reptile from North Africa. On warm sunny days some individuals may be seen basking on tree branches. Two species of geckoes, the Turkish gecko *Hemidactylus turcicus* and the Moorish gecko *Tarentula mauritanica*, and the Maltese Wall Lizard *Podarcis filfolensis* are common. Four species of snakes are known in Malta and at Għar Dalam two have been recorded, namely the Western Whip Snake *Coluber viridiflavus* and the Leopard Snake *Elaphe situla*. The only carnivore in Malta, the weasel *Mustela nivalis*, is occasionally seen running along the rubble walls, but it is usually a nocturnal hunter. Dusk brings out the bats and the Soprano Pipistrelle *Pipistrellus pygmaeus*, the Grey Long-Eared Bat *Plecotus austriacus*, and the Maghrebian Mouse-Eared Bat *Myotis punicus* hunt over the valley and fields. Birds are also present and the chattering call of the Sardinian Warbler *Sylvia melanocephala* can be heard in the low trees and bushes in the garden. A loud burst of song from the valley announces the presence of the secretive Cetti's Warbler *Cettia cetti*, while the zip-zip call of the male Zitting Cisticola *Cisticola juncidis* warns off any contenders. Robins *Erithacus rubecula* and Meadow Pipits *Anthus pratensis* are common in winter, while various other birds can sometimes be observed during migration.

Other interesting features in Għar Dalam area

Apart from the wonders of the cave, Wied Dalam provides other interesting features, some of which date back to the Bronze Age c.1500 BC. This is the fortified city at Borg in-Nadur. An impressive, massive wall of about 4.5m high is still standing there today. In the nearby fields, huts were exposed during the 1880 and 1959 excavations. It is assumed that there must be many more huts beneath the soils today. In the vicinity, a Punic tomb c.500 BC has also survived.

At Ta' Kaccatura there is the largest Roman cistern found on the islands. Columns of large blocks of stones hold a series of stone slabs which support the cistern's roof.

In the area, one can also see a pill box, which was used during the Second World War and also a watch tower dating back to the 1800s.

Today, parts of Wied Dalam are used as a fuel depot by Enemalta where fuel, mostly aviation fuel, is stored in natural as well as man-made caves. The tunnels in this area are connected to Has-Saptan (near the airport) and to Marsa (in the harbour).

NOTES

1   *M[useum] A[nnual] R*, 1936-37,17.
2   *MAR, 1931-32*, Appendix B.
3   *MAR, 1927-28*, 15.
4   *MAR, 1921-22*, 3.
5   *MAR, 1927-28*, 15.
6   *MAR, 1928-29*, 7.
7   *MAR, 1930-31*, 9.
8   *MAR, 1937*, 16-22.
9   The Piltdown Man Forgery occurred when unknown person or persons put a cranium, a mandible, and a canine tooth together to make palaeontologists believe that they all belonged to man. This confused scientists studying human evolution. However, fluorine test revealed that the cranium belonged to man in the Upper Pleistocene (about 50,000 years ago), while the mandible and canine tooth belonged to modern apes. Another test, that of nitrogen content, has produced the same result. Distinguished palaeontologists and archaeologists who took part in the excavations at Piltdown were victims of a carefully-prepared hoax.
10   A. Mifsud and S. Mifsud, *Dossier Malta – Evidence for the Magdalenian* (Malta, 1997).

FURTHER READING

Abela, G.F., *Della Descrittione di Malta isola nel mare Siciliano, con le sue antichità, ed altre notitie* (Malta, 1647).

Agius, A.J., *Għar Dalam Caves Malta - Guide Book* (Malta, 1970).

Baldacchino, A.E. & D.T. Stevens, *Is-Siġar Maltin – L-Użu u l-importanza tagħhom*, (Malta, 2000).

Bates, Dorothea M.A., 1935, *Two New Mammals from the Pleistocene of Malta, with Notes on the Associated Fauna.*

Borg, J.J., 'Għar Dalam Cave – A shelter for WWII refugees and military fuel supplies', *Treasures of Malta* (Summer 2005).

Caloi Lucia, Tassos Kotsakis, Maria R Palombo, and Carmelo Petronio, *The Pleistocene dwarf elephants of Mediterranean Islands. The Proboscidea Evolution and Palacoecology of Elephants and their relatives.* Oxford Science Publications (UK, 1996)

Cooke, John H., 1893, *The Għar Dalam Cavern, Malta and its Fossiliferous contents* (Malta, 1893).

Cooke, John H., 'On the Occurrence of Ursus Ferox in the Pleistocene of Malta', *The Geological Magazine*, Dec. III, x (1893), 344.

Despott, G., 'Excavations Conducted at Għar Dalam, Malta', *Journal of the Royal Anthropological Institute*, xlviii (1917), 214-21.

Despott, G., 'Excavations at Għar Dalam (Dalam Cave), Malta', *Journal of the Royal Anthropological Institute*, liii (January-June 1923), 18-35.

Lanfranco, S., *L-Ambjent Naturali tal-Gżejjer Maltin* (Malta, 2002).

Morana, M., *The Prehistoric Cave of Għar Dalam* (Malta, 1987).

Museum Annual Reports: 1921-22, 1927-28, 1928-29, 1930-31, 1931-32, 1936-37.

Mifsud, A. and S. Mifsud, *Dossier Malta: Evidence for the Magdalenian* (Malta, 1997).

Pedley M., M.H. Clarke, and P. Galea, *Limestone Isles in a Crystal Sea: The Geology of the Maltese Islands* (Malta, 2002).

Sultana J. and V. Falzon, *Wildlife of the Maltese Islands* (Malta, 2002).

Thomas D.S.G. & A. Goudie, A., *The Dictionary of Physical Geography*, Third edition. Blackwell (UK, 2002)

The Times [of Malta], 28 March 1986.

Trump D.H., *Malta: An Archaeological Guide* (Malta, 2000).

Trump, D.H., *Malta: Prehistory and Temples* (Malta, 2002).

Waugh, D., *Geography: An Integrated Approach*, Second edition (London, 1995).

Zammit Maempel, G., *Għar Dalam: Cave and Deposits* (Malta, 1989)

Zammit Maempel, G., *Skeletal Pathology and congenital variations in the Maltese Pleistocene hippopotamus* (London, 1993).